Detox Genes
The Pathway to Cancer

I0482303

By: Dr. L.V.K.S. Bhaskar

Contents

PREFACE

Biotransformation is the metabolic conversion of endogenous and xenobiotic chemicals to more water-soluble compounds. Xenobiotic biotransformation is accomplished by a limited number of enzymes with broad substrate specificities. Liver is considered to be the main center of detoxification. The result of biotransformation in most cases is detoxification; nevertheless, metabolism of some xenobiotics produces metabolites more reactive than their substrate compound and leads to harmful changes. The biotransformation system involves several enzyme systems that are commonly divided into two phases. Phase I enzymes are responsible for oxidation, reduction or hydrolysis and can be either detoxifying or activating. Phase II enzymes exert mainly detoxifying potential by conjugation. The cytochrome P450 enzyme superfamily, including *CYP1A1* and *CYP2E1* constitutes the majority of Phase I enzymes, while the microsomal epoxide hydrolase (mEH), *N*-acetyl transferases (NATs) and Glutathione S-Transferases are phase II enzymes primarily responsible for detoxification of xenobiotics. Accumulating data suggest that genetic polymorphisms in genes controlling carcinogen metabolism is responsible for the individual variations in cancer risk. From the broad literature available, the current book project summarizes the role specific polymorphisms in genes encoding xenobiotic metabolism, in predisposing the individuals to various cancers.

Chapter 1

Detoxification pathway

Biotransformation is the metabolic conversion of endogenous and xenobiotic chemicals to more water-soluble compounds [1,2]. The liver is one of the most important organs in the body that detoxifying foreign substances or toxins but a great amount of detoxification occurs in the gastrointestinal tract as well [3].

The detoxifying enzymes are highly polymorphic exhibiting wide phenotypic variation. Impaired ability to remove reactive substances from the body may lead to chronic disease conditions [1]. Xenobiotic metabolising enzymes are critical components in removing or detoxifying reactive metabolites of xenobiotics which make these enzyme candidates as risk factors for various disease [4]. The outcome of biotransformation in most cases is detoxification; nevertheless, metabolism of some xenobiotics produces metabolites that are more reactive than their substrate compound.

The biotransformation system involves several enzyme systems that are commonly divided into phase I and phase II (Figure 1). The phase I enzymes are responsible for oxidation, reduction or hydrolysis and can be either detoxifying or

1

activating [3]. The phase II enzymes exert mainly detoxifying potential by conjugation [1].

The cytochrome P450 enzyme superfamily, including *CYP1A1* and *CYP2E1* constitutes the majority of Phase I enzymes, while the microsomal epoxide hydrolase (mEH), *N*-acetyltransferases (NATs) and glutathione S-transferases (GSTs) are phase II enzymes primarily responsible for detoxification of xenobiotics.

Figure 1: Schematic representation of detoxification process and enzymes.

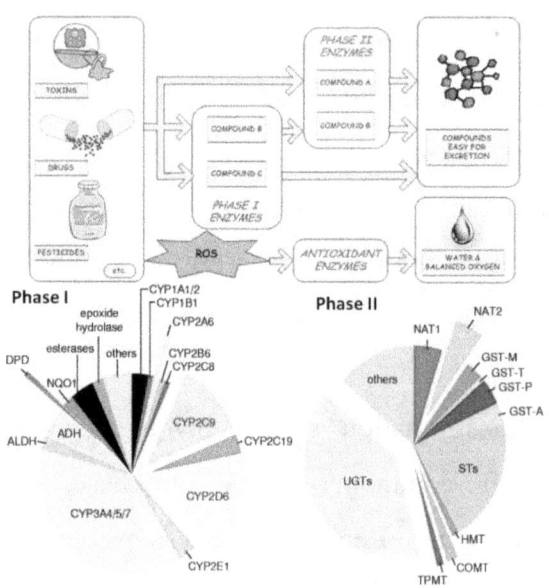

Gene polymorphisms are the basis for the huge diversity in the human phenotype, and thus the huge range of sensitivity to common as well as complex diseases, especially cancer. It is generally accepted that genomic instability is an important indicator of malignancy. Cancer is a disease of disordered gene expression due to mutations within genes or closely linked DNA that regulates the activity of those genes. By studying gene polymorphisms it is expected to reveal which changes are likely to result in disease. Single nucleotide polymorphisms (SNPs) that harbor coding and non-coding regions of a gene have the potential to change protein function or its expression and led directly to cause variations in the expression of disease phenotype.

Accumulating data suggest that genetic polymorphisms in genes that encode metabolic enzymes, carriers or receptors can affect the drug pharmacokinetics and pharmacodinamics leading to undesired therapeutic effects.

Chapter 2

The CYP450 superfamily

The CYP450 proteins are clustered into families and subfamilies according to the similarity between the amino acid sequences: where family members have $\geq$ 40% identity in amino acid sequence, members of the same subfamily share $\geq$ 55% identity [5]. The CYP450s are responsible for the metabolization of several endogenous substrates and the synthesis of hydrophobic lipids such as cholesterol, steroid hormones, bile acids and fatty acids (Figure 2). Moreover, some enzymes of P450 complex metabolize exogenous substances including drugs, environmental chemicals and pollutants as well as products derived from plants.

The metabolism of exogenous substances by CYP450 usually results in detoxification of the xenobiotics; however, the reactions triggered by such enzymes can lead to generation of toxic metabolites that contribute to the increased risk of developing cancers and other toxic effects [6]. The complete sequencing of the human genome revealed the presence of about 115 genes of CYP450, including 57 active genes and 58 pseudo-genes [5].

Figure 2: Human Hepatic CYPs

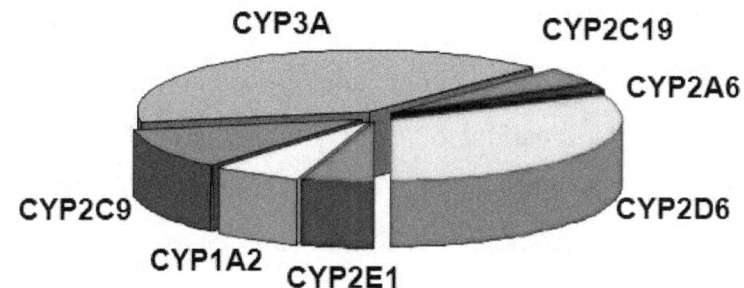

CYP1A1 and cancer proneness

CYP1A1 is a phase I detoxifying enzyme which encodes a member of the cytochrome P450 superfamily. It catalyzes the conversion of environmental procarcinogens to reactive, carcinogenic intermediates (Figure 3). *CYP1A1* mRNA is expressed in most human tissues, e.g., Lung, esophagus, stomach, small intestine, colon, prostate and breast [7,8].

Basal CYP1A1 protein expression in all tissues is thought to be low [9], but varying levels of CYP1A1 mRNA have been detected following induction by polycyclic aromatic hydrocarbons (PAHs) that found in cigarette smoke [10]. It is involved in the activation of PAHs into reactive epoxide metabolites [11]. CYP1A1 is capable of oxidizing benzo[a]pyrene and other PAH's to carcinogenic species [12].

The gene coding *CYP1A1* is composed of seven exons and localized on 15q22-24 [13]. The second exon starts the open reading frame, encoding a protein of 512 amino acids [14]. The *CYP1A1* polymorphisms are known to alter enzyme activity and play a major role in the etiology of several complex diseases [15,16]. Various *CYP1A1* gene polymorphisms have been found to be differentially associated with increased risk of several cancers (Table 1).

Figure 3: *CYP1A1* and AhR mediated carcinogenesis

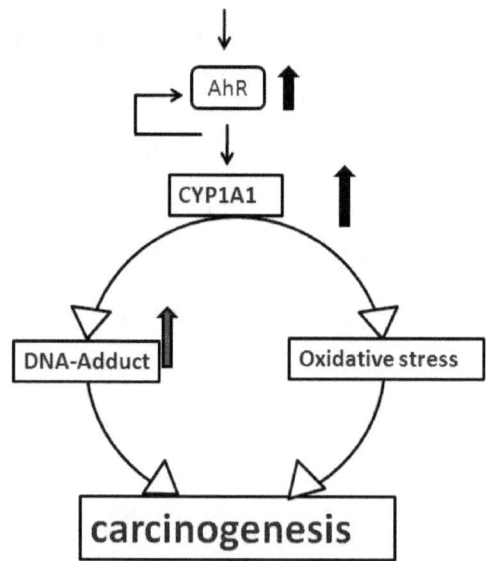

Table 1: *CYP1A1* allele Nomenclature

Nucleotide changes in Gene	Effect	Allele	References
None		*CYP1A1*1*	[17,14]
3798T>C (*MspI*)		*CYP1A1*2A*	[18]
2454A>G; 3798T>C (*MspI*)	I462V	*CYP1A1*2B*	[19]
2454A>G	I462V	*CYP1A1*2C*	[19-21]
3204T>C		*CYP1A1*3*	[22]
2452C>A	T461N	*CYP1A1*4*	[23]
2460C>A	R464S	*CYP1A1*5*	[24]
1635G>T	M331I	*CYP1A1*6*	[24]
2345_2346insT	426 Frame shift	*CYP1A1*7*	[25]
2413T>A	I448N	*CYP1A1*8*	[25]
2460C>T	R464C	*CYP1A1*9*	[25]
2499C>T	R477W	*CYP1A1*10*	[25]
2545C>G	P492R	*CYP1A1*11*	[25]

Lung cancer and *CYP1A1*

Although 90% of all lung cancer patients have a history of smoking, less than 20% of smokers only develop the disease. Even though many studies failed to show a significant association between lung cancer and *CYP1A1*2* variants, an apparent support exists for the association with at least one of them [26,27]. The 462Val (*2C*) is most frequently associated with an increased risk of lung cancer across several populations, also in subjects devoid of tobacco use [28,29]. The *2A* 3801C was first considered to be responsible for the association with lung cancer [30], but is now believed to be in linkage disequilibrium with the 462Val allele in at least some populations [31].

Esophageal cancer and *CYP1A1*

Esophageal cancer is mostly associated with excessive intake of alcohol and to some extent related to smoking as well. The association between *CYP1A1* polymorphisms and esophageal cancer risk were performed extensively in Asian individuals, the results were contradictory. However, numerous studies found 2 to 3-fold risk increase among homozygous *CYP1A1*2B* carriers [32,33], others could not demonstrate significant association for the same[34] or even found a decreased risk in persons with variant alleles [35].

CYP1A1 Msp1 T/C genotype did not influence the susceptibility of developing esophageal cancer. The MspI genotypes along with environmental exposures also did not modulate the cancer risk in North Indians [36]. Meta analysis using the 26 studies suggested that the *CYP1A1* exon7 polymorphisms may be a risk factor for esophageal cancer in Asians but not in Caucasians, whereas *CYP1A1* Msp1 was not associated with increased susceptibility to esophageal cancer [37].

Gastric cancer and *CYP1A1*

Alcohol intake, smoking, dietary compounds such as a diet high in salty and smoked foods; a diet low in fruits and vegetables are the important risk factors for gastric cancer [38].

The CYP enzymes that involved in the metabolism of these substrates and the polymorphisms in their coding genes are the important modifiers for the associations with gastric cancer. In fact the *CYP1A1* gene polymorphisms were not intensively studied in gastric cancer. A Chinese cohort study revealed 50% risk reduction in individuals carrying the *2A* variant [39]. *CYP1A1* gene polymorphisms in Japanese patients failed to show the association with gastric cancer [40]. A remarkable and statistically significant 36.5-fold increase in the risk of gastric cancer was observed among patients with *CYP1A1**2A/*2A combined with GSTM1*0/*0 Lebanese population [41].

Colorectal cancer and *CYP1A1*

The exact causes of colorectal cancer are still unknown, but certain risk factors are known to increase the chances of developing this disease. Numerous studies have associated colorectal adenoma with smoking and large bowel cancer with consumption of foods potentially containing polycyclic aromatic hydrocarbons. Eating low amounts of red meat, three servings of vegetables a day and using multivitamins with folic acid have been associated with a lowered risk of CRC.

CYP1A1 gene MspI and Ile462Val mutant genotype was significantly associated with colorectal cancer in Japanese and Hawaiians indicating that the *CYP1A1* is involved in the etiology of colorectal cancer [42]. In Caucasian *CYP1A1*2* was associated with an increased risk of colorectal cancer in both smokers and non smokers [43]. In Hungarians, the *CYP1A1*2* was overrepresented in cases but it is not associated with colorectal cancer [44]. Interestingly, two *CYP1A1* SNPs (Thr461Asn and -1738A>C) showed significant reduction in the risk of cancer [45]. A case control study from Northeast Scotland the *CYP1A1*4* (m4) variant showed a significant reduction in the risk of cancer but *CYP1A1*2A* (m1) variant is not associated with colorectal cancer [46]. In Brazilian patients the *CYP1A1*2C* (m2) "G" allele was associated with an increased risk of colorectal cancer but this polymorphism did not show any correlation between sex, grade of differentiation, stage, or evolution of the disease [47].

A recent meta-analysis revealed that *CYP1A1* Ile462Val polymorphism was significantly related with colorectal cancer risk. Subgroup ethnicity analysis showed that *CYP1A1* Ile462Val polymorphism was also significantly related with colorectal cancer risk in Europeans and Asians [48]. Another meta-analysis that examined the association between *CYP1A1* (MspI and Ile462Val) polymorphisms and risk of colorectal cancer revealed that only Ile462Val polymorphism was associated with risk of colorectal cancer. Ethnic subgroup analyses revealed that significant associations were found in Asians and Caucasians. On the contrary, *CYP1A1* MspI polymorphism does not seem capable of modifying colorectal cancer risk indicating that *CYP1A1* gene was a low-penetrance susceptibility gene in colorectal cancer development [49].

Hepatocellular cancer and *CYP1A1*

Hepatocellular carcinoma (HCC) is the most common type of liver cancer. Most cases of HCC are secondary to either a viral hepatitis infection or cirrhosis. Therefore, studies on the risk of CYP gene polymorphisms and hepatocellular carcinoma are usually performed in hepatitis-infected patients. In Taiwanese hepatitis B virus carriers, the *CYP1A1* Mspl or Ile462Val variant alleles increased the risk of HCC among smokers, but in non-smokers [50]. However, a similar study in Italian hepatitis C patients failed to replicate the similar results [51]. This could have been caused by the population stratification in the distribution of Mspl variant allele frequencies [52].

Breast cancer and *CYP1A1*

Many studies have shown that postmenopausal women who have the high levels of the estrogen estradiol in their blood have an increased risk of breast cancer. Therefore, CYPs involved in the estrogen pathway are considered as important candidate genes for the susceptibility to breast cancer. In relation to breast cancer risk only few CYP polymorphisms have been identified but these studies did not study the interaction with estrogens [53,54].

The associations between the *CYP1A1* polymorphisms and breast cancer risk have been studied extensively in the various ethnic populations with unconvincing results [55-57]. Individuals possessing *CYP1A1* Mspl variant allele exhibited an increased risk for breast cancer has been documented among

12

African-American, Indians and postmenopausal Chinese women [58,59], whereas a decreased risk in Japanese, Brazilian non-whites [54,53]. However, later studies which included a larger number of study subjects failed to show statistically significant association between *CYP1A1* genotypes and breast cancer among Chinese, Japanese and African-Americans [56,57], this is supported by a meta analysis in which the MSPI variant allele is not associated with breast cancer [60,61].

CYP1A1 461Asn allele showed increased risk of breast cancer in Caucasians of French-Canadian origin [62]. Women with *CYP1A1* 462Val variant allele in Caucasian showed statistically significant association with breast cancer [63]. A pooled meta-analysis with >9552 subjects suggests that Val462Val genotype is associated with a trend of reduced breast cancer risk, both in east-Asian women and in pre-menopausal women worldwide [60]. Recent meta-analysis points to the A2455G G allele as a risk factor for breast cancer among Caucasian subjects and did demonstrate significant associations between the MspI, T3205C and Thr461Asn polymorphisms and breast cancer [64].

Prostate cancer and *CYP1A1*

Prostate cancer is one of the most common, yet least talked about, hormone-related cancers in men. Therefore, several CYP candidate genes were studied individually as well as in combination. As tobacco-induced CYPs increased risk of prostate cancer [65,66], several studies has investigated the association between *CYP1A1* and prostate cancer. *CYP1A1* Ile462Val increased risk of prostate cancer in several populations [67-69].

The *CYP1A1* gene polymorphisms are not significantly associated with prostate cancer in Chinese [70], Turkish [71] and Brazilian population [72]. A pooled meta-analysis with 2573 subjects suggests that the *CYP1A1 MSP1* polymorphism is likely to increase the risk of sporadic prostate cancer on a wide population basis, the *Ile462Val* polymorphism may not influence this risk [73].

CYP2E1 and cancer proneness

CYP2E is a well-conserved xenobiotic-metabolizing CYP enzyme. *CYP2E1* is expressed in liver, kidney, nasal mucosa, brain, lung, and other tissues [74]. It is generally accepted, that CYP2E1 is not expressed in early fetal development; however, different time points were suggested for the beginning of

14

CYP2E1 expression [75]. With respect to postnatal CYP2E1-expression, there has been some controversy about the influence of age and sex: while earlier studies reported CYP2E1 activity in humans to be independent from age, more recent reports claimed that CYP2E1 activity rises with age [76].

Figure 4: *CYP2E1* and alcohol mediated carcinogenesis

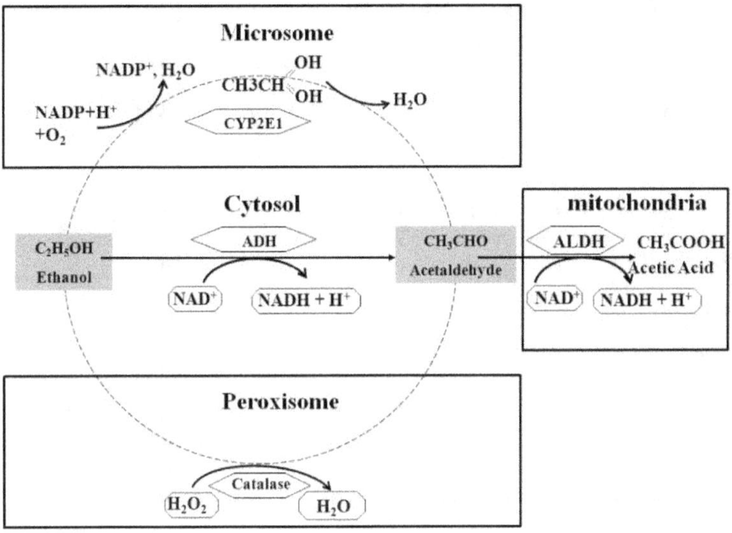

CYP2E1 is among the most conserved forms in the CYP2 family and the catalytic activities of *CYP2E1* across species are quite similar, suggesting that it has a physiological importance. CYP2E is inducible by ethanol, acetone, and other low-molecular weight substrates (Figure 4).

Table 2: Nomenclature of *CYP2E1* alleles

Allele	RFLP	Amino acid change	References
*CYP2E1*1A*			[77]
*CYP2E1*1B*	*Taq*I-		[78,79]
*CYP2E1*1C*			[80]
*CYP2E1*1D*	*Dra*I and *Xba*I		[80,81]
*CYP2E1*2*		R76H	[82]
*CYP2E1*3*		V389I	[82]
*CYP2E1*4*		V179I	[83]
*CYP2E1*5A*	*Pst*I+ *Rsa*I- *Dra*I-		[84-86]
*CYP2E1*5B*	*Pst*I+ *Rsa*I-		[84,85]
*CYP2E1*6*	*Dra*I-		[86]
*CYP2E1*7A*			[83]
*CYP2E1*7B*			[83]
*CYP2E1*7C*			[83]

The *CYP2E1* gene is mapped to chromosome 10q24.3-qter. The gene spans over 11 kb and contains 9 exons coding for a membrane-bound protein consisting of 493 amino acid residues with a molecular weight of ~ 57 kDa [87]. Both the 5'-flanking region (5'-FR) and 3'-untranslated-region (3'-UTR) harbour several mutations known to alter the transcriptional activity of the gene [88]. It is well known that the *CYP2E1* not only increase the blood concentration of acetaldehyde but also may activate these carcinogens more strongly. Activated nitrosamines have been linked to the development of numerous cancers. Various *CYP2E1* gene polymorphisms have been found to be differentially associated with increased risk of several cancers (Table 2).

Lung cancer and *CYP2E1*

Results from studies that evaluated the role of *CYP2E1* polymorphisms in relation to lung cancer have been discrepant. Half of the studies in which relations between the *CYP2E1*5 (Pst*I+ Rsa*I-/*Dra*I-) and *CYP2E1*6 (Dra*I-)* alleles and lung cancer were investigated did not find any association at all [89,90]. In the remaining studies the *CYP2E1*6 (Dra*I-)* polymorphism was associated with an increased risk of lung carcinoma [91]. On the other hand, the *CYP2E1*5B (Pst*I+ Rsa*I-)* polymorphism seems to be associated with a decreased risk of lung cancer [92]. *CYP2E1* seems to modify the effects of smoking on lung cancer, even though only a minority of studies considered the relation [93]. A recent meta analysis using 4436 cases and 6385 controls from 26 studies reported a decreased lung cancer risk among subjects carrying c1/c2 and c1/c2+c2/c2 genotypes in the Asian population and on the basis of population control in stratified analysis.

The *CYP2E1* DraI CC and CD+CC polymorphisms also showed a protective effect for lung cancer [94]. In the other meta analysis in which 21 published studies involving 9380 subjects of the association between *CYP2E1* Rsa I/Pst I polymorphism and lung cancer risk revealed both the c2 allele carriers and homozygote c2/c2 caused significant risks in Asian but not in Caucasians genetic models [95].

Esophageal cancer and *CYP2E1*

The frequency of *CYP2E1* c1/c1 genotype was significantly higher in Chinese Kazakh's patients with esophageal cancer (77.9%) than in control subjects and showed 11-fold increase of esophageal cancer risk [96], the wild-type genotype *CYP2E1* (*1/*1*) has being associated with a 3 to 5 fold increase of esophageal cancer risk in some other Chinese populations [97]. Tandem repeats in the 5'flanking region of *CYP2E1* gene, were associated with an increased risk of esophageal cancer in Japanese [89]. The *CYP2E1* variant (*6*) was also associated with an increased risk of esophageal cancer in South-African subjects [98]. On the contrary to this *CYP2E1**5B and *6 are not associated with esophageal squamous cell carcinoma (ESCC) in Brazilians [99]. A recent meta-analysis comprised of 11 published case-control studies with 1,088 cases and 2,238 controls demonstrates that *CYP2E1* Rsa I/Pst I c2 allele may be a decreased risk factor for developing esophageal cancer among Asians populations [100].

Gastric cancer and *CYP2E1*

A preliminary study in relation to gastric cancer in a Japanese population shows no association between *CYP2E1* RsaI and gastric cancer [101]. Analysis of *CYP2E1*/PstI and *CYP2E1*/DraI demonstrated the possible involvement of the *CYP2E1* polymorphism in smoking-induced gastric cancer development in Koreans [102]. Both the wild-type [40,103], as well as the *CYP2E1*5* variant [104] were mentioned as a risk factor for gastric carcinoma in homozygous individuals, but these associations were not reproduced in another study [105,41].

Colorectal cancer and *CYP2E1*

The combined variant of two polymorphisms in the untranslated region of *CYP2E1* on chromosome 10 (*2B*), related to increased risk of other cancers as well, was associated with an increased CRC risk among Hungarians [44]. However, this result could not be reproduced among Dutch Caucasians [106]. As functionality is not completely unravelled, these conflicting results are difficult to explain [107]. The *CYP2E1* c2/c2 genotype is associated with elevated odds ratio for rectal cancer, but not for colon cancer in a Chinese population [108]. Screening of *CYP2E1* RsaI and 96-bp insertion polymorphisms in 685 incident cases of colorectal cancer and 778 community controls revealed that the RsaI c2 allele is associated with a decreased risk of rectal cancer. Individuals having one or two 96-bp insertion alleles showed an increased risk of rectal cancer. Individuals with two 96-bp insertion alleles showed a 2.28-fold increase in colon cancer risk [109]. A novel *CYP2E1* locus rs1329149 was found to be significantly associated with CRC risk in Southwestern Chinese [110]. A meta-analysis based on 10 case-control studies involving 4979 colorectal cancer cases and 6012 controls revealed no association between *CYP2E1* RsaI/PstI polymorphism and colorectal cancer risk. However, in stratified analysis, Caucasians with c2c2 homozygote appeared to have an increased risk of colorectal cancer [111].

Hepatocellular cancer and *CYP2E1*

Less consistent results were found for the association between *CYP2E1* and hepatocellular cancer. The *CYP2E1* c1/c1 wild-type genotype significantly increased the risk of developing hepatocellular cancer in cigarette smokers but in those who never smoked [112]. On the contrary to this there was no increased risk of hepatocellular cancer in patients with genotypes c1/c2 and c2/c2 in Korean and Japanese [113]. In Japanese HCC patients when covariates including viremia were selected by using stepwise logistic regression analysis the frequency of *CYP2E1* C2 allele significantly higher than those of controls [114]. These results indicate a variety in allele frequencies in these Asian countries, but other explanations could hold as well.

Breast cancer and *CYP2E1*
The cellular distribution and the level of expression of *CYP2E1* assessed by immunohistochemistry, revealed that the *CYP2E1* protein is expressed in both tumour and normal breast tissue with an increased expression in breast tumours [115]. The ever drinking women with the *CYP2E1* c2 allele containing individuals had an increased risk of developing breast cancer compared to non-drinkers with the *CYP2E1* c1/c1 genotype in the Korean population [116]. *CYP2E1* PstI genotypes were not significantly different between breast cancer patients and controls living in Sousse on the middle coast of Tunisia [117].

Prostate cancer and *CYP2E1*
Analysis of *CYP2E1* gene polymorphisms in the Japanese prostate cancer patients and controls did not show the association between *CYP2E1* and Breast cancer susceptibility

[67]. Several genetic alterations have been associated with sporadic prostate cancer (PCa). The *CYP2E1* RsaI polymorphism was not statistically different between prostate cancer and controls but the DraI polymorphisms, the DD genotype is over-represented in prostate cases when compared with the control group and associated with a twofold increased risk for the development of prostate cancer Portugal population [118]. The individuals with the *CYP2E1* C1/C1 genotype and heavy smoking history showed significantly increased risk for prostate cancer in Chinese [119,120].

Chapter 3

Microsomal epoxide hydrolase

Oxidation by one or more of the phase I oxidative enzymes such as the CYP superfamily often results in the formation of a reactive xenobiotic epoxide [121]. The microsomal epoxide hydrolase (mEH) encoded by *EPHX1* is a biotransformation enzyme that metabolizes numerous reactive epoxide intermediates (Figure 5) to more water-soluble trans-dihydrodiol derivatives [121]. *EPHX1* is a smooth endoplasmic reticulum enzyme and is expressed relatively ubiquitously in most tissues and in many species [122]. It is likely that CYP and *EPHX1* enzymes cooperate via protein-protein interactions, meaning that a metabolite produced by CYP can be directly transferred to the other enzymes participating in the subsequent metabolism [123]. The *EPHX1* gene is located in chromosome 1q42.1 [124]. The gene contains nine exons, eight of which are coded [125]. The translated protein of 455 amino acids is the product of a single gene [126], although alternatively spliced non-coding regions of exon 1 have been reported [127]. The gene coding for microsomal epoxide hydrolase (*EPHX1*) exhibits Polymorphism [124]. Among these two single nucleotide polymorphisms (Tyr113His and His139Arg) have been described in the coding region of the *EPHX1* gene that produces two protein variants [124]. The expression of *EPHX1* may vary between the human tissues (Table 3). The alternative promoters are most likely defined the basis for tissue-specific expression of *EPHX1* [128], hence its expression levels are much lower in lymphocytes than in liver and lung [129].

Figure 5: Detoxification of AFBO by microsomal epoxide hydrolase.

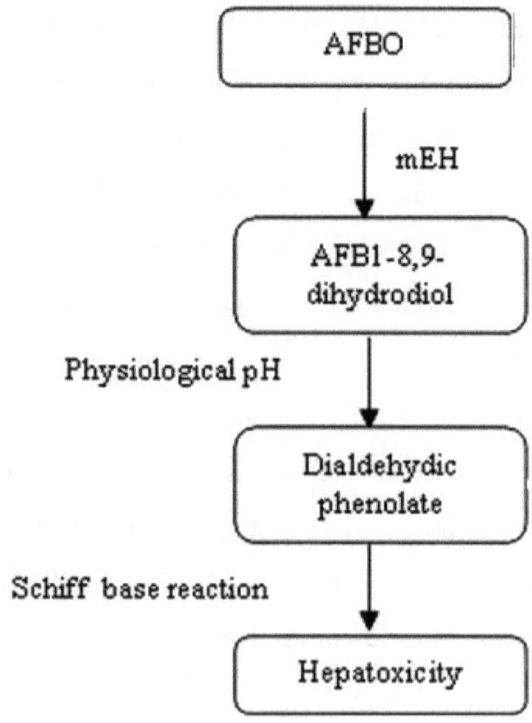

Table 3: Predicted *EPHX1* activity defined by Benhamou et al. [130].

	Y113H	H139R	Combination of Y113H and H139R
High activity		139HR, 139RR	113YH/139RR, 113YY/139HR, 113YY/139RR
Intermediate activity	113YY	139HH	113HH/139RR, 113YY/139HH, 113YH/139HR
Low activity	113YH, 113HH		113HH/139HH, 113HH/139HR, 113YH/139HH

Lung cancer and *EPHX1*

The role of *EPHX1* polymorphisms in susceptibility to lung cancer has been widely studied but the results are inconsistent [131-133]. The main reasons for the differences in results may be adopting an imperfect methodology for genotyping the exon 3 [134]. However, the earlier findings suggest a decreased risk of lung cancer for His113 slow activity allele homozygotes compared to the wild type Tyr113 allele homozygotes [135]. For exon 4 polymorphism, both increased and decreased risks have been reported. Similarly, studies on the associations between the predicted *EPHX1* activity and lung cancer risk have also given inconsistent results [136].

A systematic review and meta-analysis of 13 case-control studies revealed that H113H genotype (low activity) of *EPHX1* was associated with decreased risk while the R139R genotype (high-activity) of *EPHX1* was associated with an increased risk of lung cancer among Caucasians. Moreover, the predicted low *EPHX1* activity was associated with a modest decrease of lung cancer risk [137]. A recent comprehensive systematic review and meta-analysis of 84 studies also suggested that the predicted low *EPHX1* enzyme activity may have a potential protective effect on tobacco-related carcinogenesis of lung and

26

UADT cancers and this association is influenced by cigarette-smoking status [138].

Esophageal cancer and *EPHX1*

The predicted high mEH activity was seen more frequently in cases than controls and also the high activity genotypes of *EPHX1* were significantly increased the individual susceptibility to esophageal adenocarcinoma [35]. *EPHX1* Tyr113His polymorphism did not show significant difference in allele distribution of esophageal squamous cell carcinoma (ESCC) patients and controls in a population of North China [139]. A hospital based case control study from Taiwan suggest that the *EPHX1* His113His genotype can differentiate the association between smoking, areca chewing, and ESCC [140]. *EPHX1* gene exon 3, Tyr113His genotype was associated with higher risk of ESCC particularly at upper and middle-third anatomical locations of tumor [141].

A large-scale pathway-based candidate gene association study using 1330 single-nucleotide polymorphisms (SNPs) in 354 genes failed to show association between *EPHX1* gene polymorphisms and esophageal cancer in Caucasian [142]. A recent study from a high-incidence region of India, the Patients with the 139Arg/Arg genotype were at significantly higher risk

27

for developing a well-differentiated and moderately-differentiated grade of tumor. In contrast, the 113His/His genotype of exon 3 were a significant protective factor for esophageal cancer in tobacco smokers, betel quid chewers, and alcohol users [143].

Gastric cancer and *EPHX1*

A nested case-control study within the European Prospective Investigation into Cancer and Nutrition revealed that only the homozygous variant CC of Y113H in *EPHX1* was significantly associated with increased gastric cancer risk in ever smokers [144]. In contrast to this, polymorphisms in metabolic genes, their combination and interaction with tobacco smoke and alcohol consumption in an Italian population failed to show such a significant association between *EPHX1* polymorphisms and gastric cancer risk [145]. Recent data from a case–control study in Japan also did not observe an association between SNPs in this block and gastric cancer risk [146].

Colorectal cancer and *EPHX1*

Analysis of *EPHX1* gene polymorphisms in relation to risk of colorectal adenoma in two case-control studies nested in the Nurses' Health Study and Health Professionals Follow-up

Study cohorts did not show a significant association with overall risk of adenoma. This study also indicates that individuals exposed to > or =25 pack-years smoking were at increased risk of colorectal adenoma and that risk is related to dose of tobacco carcinogens and mEH activity level, but the results were not consistent between men and women [147].

In non-Hispanic Whites that recruited from the National Cancer Institute's Prostate, Lung, Colorectal, and Ovarian (PLCO) Cancer Screening Trial, the *EPHX1* polymorphisms are associated with increased risk of advanced colorectal adenoma, particularly among current and recent smokers [148]. In contrast to this several studies has reported that the *EPHX1* polymorphisms are not associated with sporadic colorectal cancer [106,149]. Analysis of biotransformation gene polymorphisms suggests that allelic polymorphism of metabolizing enzymes plays an important role in human colorectal carcinogenesis by affecting the metabolism of dietary carcinogens [150,151]. Meta analysis of seven studies using *EPHX1* gene polymorphisms found a weak suggestion of an antagonistic effect of *EPHX1* exon 3 low or medium metabolizer with smoking on colorectal adenoma risk [152]. In a case-only analysis, unconditional logistic regression was used to examine the associations between smoking and each SNP and between the two SNPs in nonfamilial colorectal adenoma patients, smokers with any variant allele of *EPHX1* were at increased risk for CRC, as were individuals with any variant allele of *CYP1A1* together with any variant allele of *EPHX1* [153]. A recent study of men and women with colorectal adenomas, HPs, or concurrently with both types of polyps and polyp-free controls receiving a colonoscopy failed to suggest the association between mEH genotype and colorectal polyps, nor were any statistically significant gene-environment interactions [154].

Hepatocellular cancer and *EPHX1*

In Chinese population the incidence of hepatocellular carcinoma (HCC) is found to be associated with *EPHX1* H113 allele [155]. EPHX 113HH and 139HH genotypes increased the risk of HCC, but not modify the association between peanut butter consumption indicating the unlikely role of EPHX in aflatoxin metabolism [156]. *EPHX1* 113His/His homozygotes were overrepresented in advanced stages of disease, in particular among HCC patients but these differences were more prominent among men than women. The predicted low enzyme activity was more prevalent among cirrhotic and HCC patients indicated *EPHX1* gene polymorphisms were significantly associated with HCV-related liver disease severity and HCC risk [157]. *EPHX1*, R139R imposed a risk factor for HCC and chronic hepatitis-infected subjects, the combination of *GSTM1* and T1 genotypes with either of exon 3 or 4 polymorphisms and of *EPHX1* exhibited synergistic associations for HCC development [158]. *EPHX1* gene haplotypes also exhibited sharing of a positive association with HCC risk in India [159].

Breast cancer and *EPHX1*

Analysis of *EPHX1* genotypes revealed that carriers of *EPHX1*3/*3* genotype are over-represented among breast cancer cases than in controls. The carriers of predicted low activity alleles were also exhibited higher risk of breast cancer in comparison with carriers of high *EPHX1* activity, but the results are not significant [160]. On contrary to this a significant decrease in breast cancer risk was associated with the *EPHX1* CC genotype when compared with the TT genotype in a case-control study in an Australian Caucasian population-based sample [161]. Investigation of 11 genes encoding key proteins in biosynthesis, catabolism and detoxification in breast cancer cases and controls from Germany failed to establish the relation between estrogen metabolic pathway gene polymorphisms and breast cancer risk [162]. In a subgroup of premenopausal patients with breast cancer the *EPHX1* homozygous mutant genotype has shown a significant association with the risk of breast carcinoma. The heterozygous *EPHX1* genotype was also found to be protective against breast carcinoma in the selected population [117]. The association between *EPHX1* gene polymorphisms and breast cancer was not observed in Thai women [163].

Prostate cancer and *EPHX1*

Initial studies in Israeli prostate cancer patients, the *EPHX1* His113 allele is seemingly associated with a more advanced, late onset disease [164]. Investigation of the association between prostate cancer and smoking, as well as the main and modifying effects of EPHX1 His139Arg functional polymorphisms failed to show the main effects of smoking or His139Arg polymorphisms [165]. The *EPHX1* 139Arg/Arg genotype, decreased adducts levels in both prostate tumor and nontumor prostate cells of Caucasians, but this effect was not found in African Americans [166]. Increased prostate cancer risk was observed with high, compared with no, petroleum oil/petroleum distillate in individuals carrying *EPHX1* rs17309872 [167].

Chapter 4

N-acetyltransferases

The *N*-acetyltransferases (NAT; E.C.2.3.1.5) are involved in the initial biotransformation metabolism of aromatic amines and hydrazines, and catalyses the transfer of acetyl group from acetyl CoA to the nitrogen of the substrate [168]. N-acetyltransferases are polymorphic in the population and metabolize different kinds of carcinogens that have been directly implicated in the tumor progression. *NAT2* gene (MIM # 243400) code for the NAT2 protein is located on chromosome 8p22 and spans 9.9 kb. The human *NAT2* gene has two exons with a non-coding exon at the 5′ end and an uninterrupted coding region (exon 2) of 870 nucleotides that encode a 290 amino acid protein. The *NAT2* gene is polymorphic and 66 alleles have been described till date.

Several of the *NAT2** alleles share sequence variations, and not all sequence variations would lead to change in the enzyme activity of the encoding protein (Table 4). Early genotyping studies screened for the presence of the $C^{481}T$, the $G^{590}A$, the $G^{857}A$ and sometimes the $G^{191}A$ nucleotide changes, all of which was shown to cause a slow acetylation phenotype [169]. A three fold decrease in clearance was reported between fast acetylators and slow acetylators [170].

Table 4: Nomenclature of *NAT2* alleles.

Allele	Nucleotide Change	Amino Acid Change	Phenotype	Reference
*NAT2*11A*	481C>T	L161L (synonymous)	Rapid	[171-174]
*NAT2*6B*	590G>A	Arg197Gln	Slow	[171-174]
*NAT2*7A*	857G>A	Gly286Gln	Slow	[171-176]
*NAT2*6E*	481C>T, 590G>A	L161L, R197Q	Slow	[177]

Lung cancer and *NAT2*

Initial studies examining the role of the *NAT2* polymorphic phenotype in susceptibility to lung cancer were either negative [178] or showed an insignificant overrepresentation of rapid acetylators [179,180]. Individuals with a *NAT2* rapid acetylator phenotype are thought to have an increased *NAT2* metabolic capacity and therefore have a protective role against cancer. Consistent with this hypothesis, *NAT2* rapid acetylator genotypes have been associated with a decreased risk of lung cancer, an effect that seems modulated by smoking [181,182]. In contrast, few studies failed to show association between *NAT2* acetylator genotype and lung cancer [183-185]. Never-smoking individuals with *NAT2* fast acetylator were more prone to lung cancer in Taiwan women but not in men [186].

34

NAT2 genotype responsible for slow acetylation (*NAT2*5B/*6*) was observed significantly more frequently in lung cancer patients than control subjects [187,188]. A recent meta analysis including 3945 lung cancer cases and 6085 controls from 19 published studies which were selected from 29 articles revealed very little evidence of an association between the *NAT2* polymorphism and the risk of lung cancer [189].

Esophageal cancer and *NAT2*

Till date it remains a controversy and no consensus concerning whether there is a true association between esophageal cancer and N-acetylation polymorphism. However, analysis of 71 esophageal squamous cell cancer (ESCC) patients and 329 healthy control subjects from Japan revealed overrepresentation of *NAT2* slow acetylator phenotype in esophageal cancer patients than in the controls [190]. *NAT2* slow acetylator genotype was not significantly associated with risk of esophageal cancer in studies from North India [191] and Iran [192]. In contrast to this another case control study from Kashmir Valley reported that the *NAT2* slow acetylator genotype and haplotypes increased susceptibility to ESCC [193].

Gastric cancer and *NAT2*

Previous studies showed that, individuals with *NAT2* rapid acetylators are at increased risk of developing gastric cancer in Europeans [194], and in Koreans [195]. In contrast to this, no correlation between *NAT2* polymorphic sites and gastric cancer was reported in USA [196], Poland [197], Japan [40,198] and Omani Arab population [199]. *NAT2* slow acetylator genotype is not directly associated with gastric cancer risk but may be an important modifier of the effects of environmental factors on gastric cancer risk [200]. A meta analysis using 13 studies also showed no significant association in genotype distribution between gastric cancer and control [201].

Colorectal cancer and *NAT2*

Large-scale molecular epidemiological studies that investigate the relationship between the *NAT2* varieties and colorectal cancer (CRC) are inconclusive. Several studies have shown an increased risk of CRC in patients with *NAT2* rapid acetylators [202-206]. Other studies, focused more on environmental factors acting through procarcinogenic compounds activated by *NAT2*, did not confirm an increased risk of cancer in patients with *NAT2* rapid acetylators [207-209]. Few studies of meta-analyses failed to support the hypothesis that *NAT2* alone is an important risk factor for colon cancer and suggests that *NAT2* rapid acetylation status has no specific effect on the risk of developing colon cancer [210,152].

Hepatocellular cancer and *NAT2*

A significant association between *NAT2* genetic polymorphism and hepatocellular cancer (HCC) was observed among chronic hepatitis B virus (HBV) carriers who were smokers but not among the non-smokers [211]. The smokers with a slow acetylation genotype of N-acetyltransferase 2 may be a strong risk for hepatocellular carcinoma In Chinese [212] and Germans [213]. N-acetyltransferase 2 polymorphism is not related to the risk of advanced alcoholic liver disease in Spanish individuals, but the slow acetylator genotype may predispose the ALD patients to develop HCC [214]. Although there is no association between the susceptibility of HCC and the overall *NAT2* genotypes, the individuals with rapid acetylators showed increased risk of HCC [215]. No evidence for a gene-environment interaction in HCC risk for *NAT2* genotypes was observed [216,217].

Breast cancer and *NAT2*

The association between *NAT2* acetylator phenotype or genotype and breast cancer was investigated in several studies, but the results are inconsistent. In many studies *NAT2* acetylator phenotype was not associated with breast cancer [218,219]. However, the rapid acetylator phenotype was associated with breast cancer risk [220]. Rapid-acetylation status was associated with increased risk of breast cancer both in the whole sample and among postmenopausal women [221]. Conversely, slow acetylators increased risk of breast cancer, in postmenopausal women [222]. In a meta- and pooled analysis including 13 studies, *NAT2* was not independently associated with breast cancer risk but smoking was found to be associated with increased risk in *NAT2* slow acetylators but not in rapid acetylators [223].

Prostate cancer and *NAT2*

Previous studies showed that, the slow *NAT2* genotype has been associated with a lowered prostate cancer (PC) risk while the rapid *NAT2* genotype has been associated with a non-significantly elevated PC risk [224,225]. In contrast to this *NAT2* slow acetylator genotype showed an important role in determining the risk of developing prostate cancer [226,227]. No relationship between *NAT2* genotype and prostate cancer was also observed in two studies [228,229].

Chapter 5

Glutathione S-Transferases in cancer proneness

Among the phase II carcinogen detoxifying enzymes GSTs have received great importance due to their role in detoxification of tobacco carcinogens such as PAH diol epoxides, aromatic amines, hydrazines and oxidative stress byproducts. Glutathione S-Transferases (GSTs) catalyze detoxification and activation reactions through the conjugation of biologically active electrophiles to endogenous tripeptide glutathione, predominantly in the liver [230,231]. There are 4 classes of GSTs, alpha, mu, theta and pi and all play a role in furthering the biotransformation of metabolites from phase I reactions [231]. Of note, GSTP (pi) is the only GST to be purified and cloned from the human placenta and represents 85% of the GST activity in the placenta as early as the first trimester [230].

The GST mu (GSTM) and GST theta (GSTT) classes have been studied because of their role in detoxification of activated nicotine metabolites and xenobiotics [230]. They are known for their involvement in the detoxification of epoxides created by the CYP450s [232]. There are 5 sub-classes of GSTM (*GSTM1*-5), which cluster on chromosome 1p13 [233,234]. There is a large difference in the expression of GSTM between different tissue types [231]. The most commonly expressed GSTM is *GSTM1* [231]. The *GSTM1* subclass is of particular interest because of its prominent null allele. Only about 40-60% of individuals in the population express *GSTM1* and, for those who do not express the gene, there is an increased susceptibility to DNA-adduct formation and cytogenetic damage [235,236]. GSTT has two sub-classes, designated *GSTT1* and *GSTT2* [237]. These enzymes are also found in the liver, but have widespread expression [231]. Similarly to the *GSTM1* locus, *GSTT1* has a null allele that can be found in 10-40% of individuals, depending on the population [231]. The null alleles for both *GSTM1* and *GSTT1* cause an absence of enzyme activity, and possibly increasing the amount of active metabolites in the body [231,238]. Therefore, *GSTM1* and *GSTT1* activity may impact the effects of harmful intermediates created by phase I enzymes on a fetus [230].

Various workers have investigated the function of GST variations in modulating individual risk for head and neck cancers. Polymorphic variants are frequently observed in *GSTT*1, *GSTP*1 and *GSTM*1 genes. *GSTM*1 and *GSTP*1 plays important role in the metabolization of PAH diol epoxides while *GSTT*1 participates in the detoxification of potentially carcinogenic monohalomethanes and reactive epoxide

40

metabolites of butadiene, both of which are constituents of tobacco smoke. Therefore, several workers have worked on the polymorphisms of these genes with respect to cancer risk for the past two decades [239,240]. GST polymorphisms may have prognostic value in prostate cancer [241], lung cancer [242] and non-Hodgkin's lymphoma [243]. Conflicting results between the GST polymorphisms and various cancer treatment outcomes are also reported in literature [244,239]. Carcinogen metabolism is affected by majority of polymorphisms mainly the SNPs. Loss of segments are not frequent and loss of gene as null allele is rare. Large number of studies involved screening of GST genotypic status because GST enzymes provide protection to somatic cells from DNA damage from carcinogens. Persons having homozygous deletions of *GSTT1* or *GSTM1* genes lack the ability to detoxify the carcinogen and in turn increase the risk of cancer. As most of the isoforms of GSTs do not have specificities regarding substrate therefore the loss of one isoenzyme may be overcome by other. Due to this reason genotypic status of all the GST genotypes is required in order to interpret the role of GSTs in cancer development. On the other hand, most of the studies were aimed at homozygous screening of deletion genotype and phenotype was designated as none or all. Approximately 20-30% of GSH concentration depletion can be detrimental to cellular processes due to impairment of the GSTs to attach with GSH and work against carcinogens [245]. GSTs are involved in the conversion of many groups of carcinogens through attachment with glutathione. Deletion polymorphisms of *GSTT*1 are frequent in many populations. Homozygous loss of *GSTT*1 gene removes respective GST enzyme functionality and also its ability to attach with specific respective substrate [246].

GST gene polymorphisms have been widely studied in various cancer types and recent meta-analyses have demonstrated associations with bladder, liver, lung, and kidney cancers as well as acute lymphoblastic leukaemia. The relation between GSTs genotype and cancer risk have been studied but no clear pattern has emerged [247,248]. Studies that involved one or more GSTs give contrasting results because the sample under observation and other environmental factors associated with them vary from one study group to other [247].

Cancer susceptibility and ethnicity

Cancer incidence and mortality rates vary significantly among different ethnic and racial groups. The incidence rates of esophageal cancer are approximately three times more in black Americans than the white Americans. In Black Americans the incidence of liver, cervical, stomach cancer and oral cancers and are premenopausal breast cancer higher. Whereas the incidence rates of melanoma, leukemia, lymphoma, and cancers of the endometrium, thyroid, bladder, ovary, testis, and brain, as well as postmenopausal breast cancer are higher in white Americans [249,250].

The cancer rates are generally lower in Hispanics compared to the white or black Americans [251-253]. The genetic polymorphisms affecting carcinogen metabolism and DNA repair is one of the major biologic contributors of cancer risks in certain ethnic or racial groups. The variations that observed in the serum concentration of the DDT metabolite DDE in black and white American women reflect genetic differences that affecting the breast cancer [254].

There is considerable variation in the proportion of slow acetylators in different ethnic groups. The Eskimos and Japanese have the lowest rates for slow acetylators (about 10%) with the Chinese having a rate of about 20% [255]. Slow acetylators are less common among the populations of East Asia. In one study from Berlin, 62% of the Caucasian subjects were classified as slow acetylators [256] while in France about 53% were slow acetylators [183]. These frequencies are consistent with the respective ethnic or racial differences in bladder cancer rates. Furthermore, increased incidence of breast cancer in descendants of Asian immigrants to the United States provides strong evidence that environmental factors affect cancer patterns.

References

1. Liska D, Lyon M, Jones DS (2006). Explore (NY) 2 (2):122-140
2. Dekant W (2009). EXS 99:57-86
3. Liska DJ (1998). Altern Med Rev 3 (3):187-198
4. Xu C, Li CY, Kong AN (2005). Arch Pharm Res 28 (3):249-268
5. Nelson DR (2009). Hum Genomics 4 (1):59-65
6. Gueguen Y, Mouzat K, Ferrari L, Tissandie E, Lobaccaro JM, Batt AM et al. (2006). Ann Biol Clin (Paris) 64 (6):535-548
7. Hosoya T, Harada N, Mimura J, Motohashi H, Takahashi S, Nakajima O et al. (2008). Biochem Biophys Res Commun 365 (3):562-567
8. Dey A, Jones JE, Nebert DW (1999). Biochem Pharmacol 58 (3):525-537
9. Nebert DW, Dalton TP, Okey AB, Gonzalez FJ (2004). J Biol Chem 279 (23):23847-23850
10. Roos PH, Bolt HM (2005). Expert Opin Drug Metab Toxicol 1 (2):187-202
11. Kleiner HE, Vulimiri SV, Hatten WB, Reed MJ, Nebert DW, Jefcoate CR et al. (2004). Chem Res Toxicol 17 (12):1667-1674
12. Sparfel L, Huc L, Le Vee M, Desille M, Lagadic-Gossmann D, Fardel O (2004). Biochem Pharmacol 67 (9):1711-1719
13. Hildebrand CE, Gonzalez FJ, McBride OW, Nebert DW (1985). Nucleic Acids Res 13 (6):2009-2016
14. Kawajiri K, Watanabe J, Gotoh O, Tagashira Y, Sogawa K, Fujii-Kuriyama Y (1986). Eur J Biochem 159 (2):219-225
15. Kristiansen W, Haugen TB, Witczak O, Andersen JM, Fossa SD, Aschim EL (2011). Int J Androl 34 (1):77-83
16. Wright CM, Larsen JE, Colosimo ML, Barr JJ, Chen L, McLachlan RE et al. (2010). Eur Respir J 35 (1):152-159
17. Jaiswal AK, Gonzalez FJ, Nebert DW (1985). Nucleic Acids Res 13 (12):4503-4520

18. Spurr NK, Gough AC, Stevenson K, Wolf CR (1987). Nucleic Acids Res 15 (14):5901

19. Hayashi S, Watanabe J, Nakachi K, Kawajiri K (1991). J Biochem 110 (3):407-411

20. Zhang ZY, Fasco MJ, Huang L, Guengerich FP, Kaminsky LS (1996). Cancer Res 56 (17):3926-3933

21. Persson I, Johansson I, Ingelman-Sundberg M (1997). Biochem Biophys Res Commun 231 (1):227-230

22. Crofts F, Cosma GN, Currie D, Taioli E, Toniolo P, Garte SJ (1993). Carcinogenesis 14 (9):1729-1731

23. Cascorbi I, Brockmoller J, Roots I (1996). Cancer Res 56 (21):4965-4969

24. Chevalier D, Allorge D, Lo-Guidice JM, Cauffiez C, Lhermitte M, Lafitte JJ et al. (2001). Hum Mutat 17 (4):355

25. Saito T, Egashira M, Kiyotani K, Fujieda M, Yamazaki H, Kiyohara C et al. (2003). Drug Metab Pharmacokinet 18 (3):218-221

26. London SJ, Yuan JM, Coetzee GA, Gao YT, Ross RK, Yu MC (2000). Cancer Epidemiol Biomarkers Prev 9 (9):987-991

27. Chen S, Xue K, Xu L, Ma G, Wu J (2001). Mutat Res 458 (1-2):41-47

28. Song N, Tan W, Xing D, Lin D (2001). Carcinogenesis 22 (1):11-16

29. Ozturk O, Isbir T, Yaylim I, Kocaturk CI, Gurses A (2003). In Vivo 17 (6):625-632

30. Sreelekha TT, Rajesh M, Anil Kumar V, Madhavan J, Balaram P (2010). Mol Med Report 3 (6):971-976

31. Yuan X, Zhou G, Zhai Y, Xie W, Cui Y, Cao J et al. (2008). Cancer Epidemiol Biomarkers Prev 17 (12):3621-3627

32. Wu MT, Lee JM, Wu DC, Ho CK, Wang YT, Lee YC et al. (2002). Br J Cancer 87 (5):529-532

33. Wang AH, Sun CS, Li LS, Huang JY, Chen QS, Xu DZ (2004). World J Gastroenterol 10 (7):940-944

34. Abbas A, Delvinquiere K, Lechevrel M, Lebailly P, Gauduchon P, Launoy G et al. (2004). World J Gastroenterol 10 (23):3389-3393

35. Wang LD, Zheng S, Liu B, Zhou JX, Li YJ, Li JX (2003). World J Gastroenterol 9 (7):1394-1397

36. Jain M, Kumar S, Ghoshal UC, Mittal B (2007). Oncol Res 16 (9):437-443

37. Zhuo WL, Zhang YS, Wang Y, Zhuo XL, Zhu B, Cai L et al. (2009). Arch Med Res 40 (3):169-179

38. Lazarevic K, Nagorni A, Rancic N, Milutinovic S, Stosic L, Ilijev I (2010). J BUON 15 (1):89-93

39. Roth MJ, Abnet CC, Johnson LL, Mark SD, Dong ZW, Taylor PR et al. (2004). Cancer Causes Control 15 (10):1077-1083

40. Suzuki S, Muroishi Y, Nakanishi I, Oda Y (2004). J Gastroenterol 39 (3):220-230

41. Darazy M, Balbaa M, Mugharbil A, Saeed H, Sidani H, Abdel-Razzak Z (2011). Genet Test Mol Biomarkers 15 (6):423-429

42. Sivaraman L, Leatham MP, Yee J, Wilkens LR, Lau AF, Le Marchand L (1994). Cancer Res 54 (14):3692-3695

43. Slattery ML, Samowtiz W, Ma K, Murtaugh M, Sweeney C, Levin TR et al. (2004). Am J Epidemiol 160 (9):842-852

44. Kiss I, Sandor J, Pajkos G, Bogner B, Hegedus G, Ember I (2000). Anticancer Res 20 (1B):519-522

45. Landi S, Gemignani F, Moreno V, Gioia-Patricola L, Chabrier A, Guino E et al. (2005). Pharmacogenet Genomics 15 (8):535-546

46. Little J, Sharp L, Masson LF, Brockton NT, Cotton SC, Haites NE et al. (2006). Int J Cancer 119 (9):2155-2164

47. Pereira Serafim PV, Cotrim Guerreiro da Silva ID, Manoukias Forones N (2008). Int J Biol Markers 23 (1):18-23

48. Jin JQ, Hu YY, Niu YM, Yang GL, Wu YY, Leng WD et al. (2011). World J Gastroenterol 17 (2):260-266

49. Zheng Y, Wang JJ, Sun L, Li HL (2011). Mol Biol Rep

50. Yu MW, Chiu YH, Yang SY, Santella RM, Chern HD, Liaw YF et al. (1999). Br J Cancer 80 (3-4):598-603

51. Silvestri L, Sonzogni L, De Silvestri A, Gritti C, Foti L, Zavaglia C et al. (2003). Int J Cancer 104 (3):310-317

52. Vineis P (2002). Toxicology 181-182:457-462

53. Hefler LA, Tempfer CB, Grimm C, Lebrecht A, Ulbrich E, Heinze G et al. (2004). Cancer 101 (2):264-269

54. Miyoshi Y, Takahashi Y, Egawa C, Noguchi S (2002). Breast J 8 (4):209-215

55. Masson LF, Sharp L, Cotton SC, Little J (2005). Am J Epidemiol 161 (10):901-915

56. Miyoshi Y, Ando A, Hasegawa S, Ishitobi M, Yamamura J, Irahara N et al. (2003). Eur J Cancer 39 (17):2531-2537

57. Boyapati SM, Shu XO, Gao YT, Cai Q, Jin F, Zheng W (2005). Cancer 103 (11):2228-2235

58. Moysich KB, Shields PG, Freudenheim JL, Schisterman EF, Vena JE, Kostyniak P et al. (1999). Cancer Epidemiol Biomarkers Prev 8 (1):41-44

59. Chacko P, Joseph T, Mathew BS, Rajan B, Pillai MR (2005). Mutat Res 581 (1-2):153-163

60. Chen C, Huang Y, Li Y, Mao Y, Xie Y (2007). J Hum Genet 52 (5):423-435

61. Sergentanis TN, Economopoulos KP (2010). Breast Cancer Res Treat 122 (2):459-469

62. Krajinovic M, Ghadirian P, Richer C, Sinnett H, Gandini S, Perret C et al. (2001). Int J Cancer 92 (2):220-225

63. Zhang Y, Wise JP, Holford TR, Xie H, Boyle P, Zahm SH et al. (2004). Am J Epidemiol 160 (12):1177-1183

64. Sergentanis TN, Economopoulos KP (2010). Breast Cancer Res Treat

65. Suzuki K, Matsui H, Nakazato H, Koike H, Okugi H, Hasumi M et al. (2003). Cancer Lett 195 (2):177-183

66. Sterling KM, Jr., Cutroneo KR (2004). J Cell Biochem 91 (2):423-429

67. Murata M, Watanabe M, Yamanaka M, Kubota Y, Ito H, Nagao M et al. (2001). Cancer Lett 165 (2):171-177

68. Chang BL, Zheng SL, Isaacs SD, Turner A, Hawkins GA, Wiley KE et al. (2003). Int J Cancer 106 (3):375-378

69. Aktas D, Hascicek M, Sozen S, Ozen H, Tuncbilek E (2004). Cancer Genet Cytogenet 154 (1):81-85

70. Li M, Guan TY, Li Y, Na YQ (2008). Chin Med J (Engl) 121 (4):305-308

71. Silig Y, Pinarbasi H, Gunes S, Ayan S, Bagci H, Cetinkaya O (2006). Cancer Invest 24 (1):41-45

72. Rodrigues IS, Kuasne H, Losi-Guembarovski R, Fuganti PE, Gregorio EP, Kishima MO et al. (2010). Urol Oncol

73. Shaik AP, Jamil K, Das P (2009). Urol J 6 (2):78-86

74. Sohda T, Shimizu M, Kamimura S, Okumura M (1993). Alcohol Alcohol Suppl 1B:69-75

75. Johnsrud EK, Koukouritaki SB, Divakaran K, Brunengraber LL, Hines RN, McCarver DG (2003). J Pharmacol Exp Ther 307 (1):402-407

76. Bebia Z, Buch SC, Wilson JW, Frye RF, Romkes M, Cecchetti A et al. (2004). Clin Pharmacol Ther 76 (6):618-627

77. Umeno M, McBride OW, Yang CS, Gelboin HV, Gonzalez FJ (1988). Biochemistry 27 (25):9006-9013

78. McBride OW, Umeno M, Gelboin HV, Gonzalez FJ (1987). Nucleic Acids Res 15 (23):10071

79. Brockmoller J, Cascorbi I, Kerb R, Roots I (1996). Cancer Res 56 (17):3915-3925

80. Hu Y, Hakkola J, Oscarson M, Ingelman-Sundberg M (1999). Biochem Biophys Res Commun 263 (2):286-293

81. McCarver DG, Byun R, Hines RN, Hichme M, Wegenek W (1998). Toxicol Appl Pharmacol 152 (1):276-281

82. Hu Y, Oscarson M, Johansson I, Yue QY, Dahl ML, Tabone M et al. (1997). Mol Pharmacol 51 (3):370-376

83. Fairbrother KS, Grove J, de Waziers I, Steimel DT, Day CP, Crespi CL et al. (1998). Pharmacogenetics 8 (6):543-552

84. Watanabe J, Hayashi S, Nakachi K, Imai K, Suda Y, Sekine T et al. (1990). Nucleic Acids Res 18 (23):7194

85. Hayashi S, Watanabe J, Kawajiri K (1991). J Biochem 110 (4):559-565

86. Persson I, Johansson I, Bergling H, Dahl ML, Seidegard J, Rylander R et al. (1993). FEBS Lett 319 (3):207-211

87. Lewis DF, Bird MG, Parke DV (1997). Toxicology 118 (2-3):93-113

88. Chen JM, Ferec C, Cooper DN (2006). Hum Genet 120 (1):1-21

89. Itoga S, Nomura F, Makino Y, Tomonaga T, Shimada H, Ochiai T et al. (2002). Alcohol Clin Exp Res 26 (8 Suppl):15S-19S

90. Sobti RC, Sharma S, Joshi A, Jindal SK, Janmeja A (2004). Mol Cell Biochem 266 (1-2):1-9

91. Wu X, Amos CI, Kemp BL, Shi H, Jiang H, Wan Y et al. (1998). Cancer Epidemiol Biomarkers Prev 7 (1):13-18

92. Quinones L, Lucas D, Godoy J, Caceres D, Berthou F, Varela N et al. (2001). Cancer Lett 174 (1):35-44

93. Eom SY, Zhang YW, Kim SH, Choe KH, Lee KY, Park JD et al. (2009). Cancer Causes Control 20 (2):137-145

94. Wang Y, Yang H, Li L, Wang H, Zhang C, Yin G et al. (2010). Eur J Cancer 46 (4):758-764

95. Zhan P, Wang J, Zhang Y, Qiu LX, Zhao SF, Qian Q et al. (2010). Lung Cancer 69 (1):19-25

96. Lu XM, Zhang YM, Lin RY, Arzi G, Wang X, Zhang YL et al. (2005). World J Gastroenterol 11 (24):3651-3654

97. Tan W, Song N, Wang GQ, Liu Q, Tang HJ, Kadlubar FF et al. (2000). Cancer Epidemiol Biomarkers Prev 9 (6):551-556

98. Li D, Dandara C, Parker MI (2005). Clin Chem Lab Med 43 (4):370-375

99. Rossini A, Rapozo DC, Soares Lima SC, Guimaraes DP, Ferreira MA, Teixeira R et al. (2007). Carcinogenesis 28 (12):2537-2542

100. Niu Y, Yuan H, Leng W, Pang Y, Gu N, Chen N (2011). Med Oncol 28 (1):182-187

101. Kato S, Onda M, Matsukura N, Tokunaga A, Tajiri T, Kim DY et al. (1995). Pharmacogenetics 5 Spec No:S141-144

102. Park GT, Lee OY, Kwon SJ, Lee CG, Yoon BC, Hahm JS et al. (2003). J Gastroenterol Hepatol 18 (11):1257-1263

103. Cai L, Zheng ZL, Zhang ZF (2005). World J Gastroenterol 11 (12):1867-1871

104. Wu MS, Chen CJ, Lin MT, Wang HP, Shun CT, Sheu JC et al. (2002). Int J Colorectal Dis 17 (5):338-343

105. Masuda G, Tokunaga A, Shirakawa T, Togashi A, Kiyama T, Kato S et al. (2007). Gastric Cancer 10 (2):98-103

106. van der Logt EM, Bergevoet SM, Roelofs HM, Te Morsche RH, Dijk Y, Wobbes T et al. (2006). Mutat Res 593 (1-2):39-49

107. Kato S, Shields PG, Caporaso NE, Sugimura H, Trivers GE, Tucker MA et al. (1994). Cancer Epidemiol Biomarkers Prev 3 (6):515-518

108. Gao CM, Takezaki T, Wu JZ, Chen MB, Liu YT, Ding JH et al. (2007). World J Gastroenterol 13 (43):5725-5730

109. Morita M, Le Marchand L, Kono S, Yin G, Toyomura K, Nagano J et al. (2009). Cancer Epidemiol Biomarkers Prev 18 (1):235-241

110. Yang H, Zhou Y, Zhou Z, Liu J, Yuan X, Matsuo K et al. (2009). Cancer Epidemiol Biomarkers Prev 18 (9):2522-2527

111. Zhou GW, Hu J, Li Q (2010). World J Gastroenterol 16 (23):2949-2953

112. Yu MW, Gladek-Yarborough A, Chiamprasert S, Santella RM, Liaw YF, Chen CJ (1995). Gastroenterology 109 (4):1266-1273

113. Lee HS, Yoon JH, Kamimura S, Iwata K, Watanabe H, Kim CY (1997). Int J Cancer 71 (5):737-740

114. Munaka M, Kohshi K, Kawamoto T, Takasawa S, Nagata N, Itoh H et al. (2003). J Cancer Res Clin Oncol 129 (6):355-360

115. Kapucuoglu N, Coban T, Raunio H, Pelkonen O, Edwards RJ, Boobis AR et al. (2003). Cancer Lett 196 (2):153-159

116. Choi JY, Abel J, Neuhaus T, Ko Y, Harth V, Hamajima N et al. (2003). Pharmacogenetics 13 (2):67-72

117. Khedhaier A, Hassen E, Bouaouina N, Gabbouj S, Ahmed SB, Chouchane L (2008). BMC Cancer 8:109

118. Ferreira PM, Medeiros R, Vasconcelos A, Costa S, Pinto D, Morais A et al. (2003). Eur J Cancer Prev 12 (3):205-211

119. Yang J, Wu HF, Zhang W, Gu M, Hua LX, Sui YG et al. (2006). Front Biosci 11:2052-2060

120. Yang J, Gu M, Song NH, Feng NH, Hua LX, Ju XB et al. (2009). Zhonghua Nan Ke Xue 15 (1):7-11

121. Decker M, Arand M, Cronin A (2009). Arch Toxicol 83 (4):297-318

122. Coller JK, Fritz P, Zanger UM, Siegle I, Eichelbaum M, Kroemer HK et al. (2001). Histochem J 33 (6):329-336

123. Taura Ki K, Yamada H, Naito E, Ariyoshi N, Mori Ma MA, Oguri K (2002). Arch Biochem Biophys 402 (2):275-280

124. Hartsfield JK, Jr., Sutcliffe MJ, Everett ET, Hassett C, Omiecinski CJ, Saari JA (1998). Cytogenet Cell Genet 83 (1-2):44-45

125. Hassett C, Robinson KB, Beck NB, Omiecinski CJ (1994). Genomics 23 (2):433-442

126. Skoda RC, Demierre A, McBride OW, Gonzalez FJ, Meyer UA (1988). J Biol Chem 263 (3):1549-1554

127. Gaedigk A, Leeder JS, Grant DM (1997). DNA Cell Biol 16 (11):1257-1266

128. Liang SH, Hassett C, Omiecinski CJ (2005). Mol Pharmacol 67 (1):220-230

129. Omiecinski CJ, Aicher L, Holubkov R, Checkoway H (1993). Pharmacogenetics 3 (3):150-158

130. Benhamou S, Reinikainen M, Bouchardy C, Dayer P, Hirvonen A (1998). Cancer Res 58 (23):5291-5293

131. Erkisi Z, Yaylim-Eraltan I, Turna A, Gormus U, Camlica H, Isbir T (2010). Tumori 96 (5):756-763

132. Graziano C, Comin CE, Crisci C, Novelli L, Politi L, Messerini L et al. (2009). Lung Cancer 63 (2):187-193

133. Sun XW, Ma YY, Wang B (2007). Zhonghua Yu Fang Yi Xue Za Zhi 41 Suppl:30-34

134. Yoshikawa M, Hiyama K, Ishioka S, Maeda H, Maeda A, Yamakido M (2000). Int J Mol Med 5 (1):49-53

135. Gsur A, Zidek T, Schnattinger K, Feik E, Haidinger G, Hollaus P et al. (2003). Br J Cancer 89 (4):702-706

136. Cajas-Salazar N, Au WW, Zwischenberger JB, Sierra-Torres CH, Salama SA, Alpard SK et al. (2003). Cancer Genet Cytogenet 145 (2):97-102

137. Kiyohara C, Yoshimasu K, Takayama K, Nakanishi Y (2006). Epidemiology 17 (1):89-99

138. Li X, Hu Z, Qu X, Zhu J, Li L, Ring BZ et al. (2011). PLoS One 6 (3):e14749

139. Zhang JH, Jin X, Li Y, Wang R, Guo W, Wang N et al. (2003). World J Gastroenterol 9 (12):2654-2657

140. Lin YC, Wu DC, Lee JM, Hsu HK, Kao EL, Yang CH et al. (2006). Cancer Lett 237 (2):281-288

141. Jain M, Tilak AR, Upadhyay R, Kumar A, Mittal B (2008). Toxicol Appl Pharmacol 230 (2):247-251

142. Liu CY, Wu MC, Chen F, Ter-Minassian M, Asomaning K, Zhai R et al. (2010). Carcinogenesis 31 (7):1259-1263

143. Ihsan R, Chattopadhyay I, Phukan R, Mishra AK, Purkayastha J, Sharma J et al. (2010). J Gastroenterol Hepatol 25 (8):1456-1462

144. Agudo A, Sala N, Pera G, Capella G, Berenguer A, Garcia N et al. (2006). Cancer Epidemiol Biomarkers Prev 15 (12):2427-2434

145. Boccia S, Sayed-Tabatabaei FA, Persiani R, Gianfagna F, Rausei S, Arzani D et al. (2007). BMC Cancer 7:206

146. Ikeda S, Sasazuki S, Natsukawa S, Shaura K, Koizumi Y, Kasuga Y et al. (2008). Am J Gastroenterol 103 (6):1476-1487

147. Tranah GJ, Giovannucci E, Ma J, Fuchs C, Hankinson SE, Hunter DJ (2004). Carcinogenesis 25 (7):1211-1218

148. Huang WY, Chatterjee N, Chanock S, Dean M, Yeager M, Schoen RE et al. (2005). Cancer Epidemiol Biomarkers Prev 14 (1):152-157

149. Mitrou PN, Watson MA, Loktionov AS, Cardwell C, Gunter MJ, Atkin WS et al. (2007). Carcinogenesis 28 (4):875-882

150. Kiss I, Orsos Z, Gombos K, Bogner B, Csejtei A, Tibold A et al. (2007). Anticancer Res 27 (4C):2931-2937

151. Skjelbred CF, Saebo M, Hjartaker A, Grotmol T, Hansteen IL, Tveit KM et al. (2007). BMC Cancer 7:228

152. Raimondi S, Botteri E, Iodice S, Lowenfels AB, Maisonneuve P (2009). Mutat Res 670 (1-2):6-14

153. Pande M, Amos CI, Eng C, Frazier ML (2010). Mol Carcinog 49 (11):974-980

154. Burnett-Hartman AN, Newcomb PA, Mandelson MT, Adams SV, Wernli KJ, Shadman M et al. (2011). Nutr Cancer 63 (4):583-592

155. McGlynn KA, Rosvold EA, Lustbader ED, Hu Y, Clapper ML, Zhou T et al. (1995). Proc Natl Acad Sci U S A 92 (6):2384-2387

156. Kelly EJ, Erickson KE, Sengstag C, Eaton DL (2002). Toxicol Sci 65 (1):35-42

157. Sonzogni L, Silvestri L, De Silvestri A, Gritti C, Foti L, Zavaglia C et al. (2002). Hepatology 36 (1):195-201

158. Kiran M, Chawla YK, Kaur J (2008). DNA Cell Biol 27 (12):687-694

159. Kiran M, Chawla YK, Jain M, Kaur J (2009). DNA Cell Biol 28 (11):573-577

160. Sarmanova J, Susova S, Gut I, Mrhalova M, Kodet R, Adamek J et al. (2004). Eur J Hum Genet 12 (10):848-854

161. Spurdle AB, Chang JH, Byrnes GB, Chen X, Dite GS, McCredie MR et al. (2007). Cancer Epidemiol Biomarkers Prev 16 (4):769-774

162. Justenhoven C, Hamann U, Schubert F, Zapatka M, Pierl CB, Rabstein S et al. (2008). Breast Cancer Res Treat 108 (1):137-149

163. Sangrajrang S, Sato Y, Sakamoto H, Ohnami S, Laird NM, Khuhaprema T et al. (2009). Int J Cancer 125 (4):837-843

164. Figer A, Friedman T, Manguoglu AE, Flex D, Vazina A, Novikov I et al. (2003). Isr Med Assoc J 5 (10):741-745

165. Nock NL, Liu X, Cicek MS, Li L, Macarie F, Rybicki BA et al. (2006). Cancer Epidemiol Biomarkers Prev 15 (4):756-761

166. Nock NL, Tang D, Rundle A, Neslund-Dudas C, Savera AT, Bock CH et al. (2007). Cancer Epidemiol Biomarkers Prev 16 (6):1236-1245

167. Figueroa JD, Malats N, Garcia-Closas M, Real FX, Silverman D, Kogevinas M et al. (2008). Carcinogenesis 29 (10):1955-1962

168. Riddle B, Jencks WP (1971). J Biol Chem 246 (10):3250-3258

169. Hickman D, Risch A, Camilleri JP, Sim E (1992). Pharmacogenetics 2 (5):217-226

170. Dorne JL, Walton K, Renwick AG (2003). Food Chem Toxicol 41 (2):225-245

171. Delomenie C, Sica L, Grant DM, Krishnamoorthy R, Dupret JM (1996). Pharmacogenetics 6 (2):177-185

172. Fretland AJ, Leff MA, Doll MA, Hein DW (2001). Pharmacogenetics 11 (3):207-215

173. Hein DW, Fretland AJ, Doll MA (2006). Int J Cancer 119 (5):1208-1211

174. Zang Y, Doll MA, Zhao S, States JC, Hein DW (2007). Carcinogenesis 28 (8):1665-1671

175. Agundez JA, Olivera M, Martinez C, Ladero JM, Benitez J (1996). Pharmacogenetics 6 (5):423-428

176. Cascorbi I, Drakoulis N, Brockmoller J, Maurer A, Sperling K, Roots I (1995). Am J Hum Genet 57 (3):581-592

177. Dandara C, Masimirembwa CM, Magimba A, Kaaya S, Sayi J, Sommers DK et al. (2003). Pharmacogenetics 13 (1):55-58

178. Burgess EJ, Trafford JA (1985). Eur J Respir Dis 67 (1):17-19

179. Roots I, Drakoulis N, Ploch M, Heinemeyer G, Loddenkemper R, Minks T et al. (1988). Klin Wochenschr 66 Suppl 11:87-97

180. Philip PA, Fitzgerald DL, Cartwright RA, Peake MD, Rogers HJ (1988). Carcinogenesis 9 (3):491-493

181. Zhou W, Liu G, Thurston SW, Xu LL, Miller DP, Wain JC et al. (2002). Cancer Epidemiol Biomarkers Prev 11 (1):15-21

182. Sorensen M, Autrup H, Tjonneland A, Overvad K, Raaschou-Nielsen O (2005). Cancer Lett 221 (2):185-190

183. Bouchardy C, Mitrunen K, Wikman H, Husgafvel-Pursiainen K, Dayer P, Benhamou S et al. (1998). Pharmacogenetics 8 (4):291-298

184. Zienolddiny S, Campa D, Lind H, Ryberg D, Skaug V, Stangeland LB et al. (2008). Carcinogenesis 29 (6):1164-1169

185. Belogubova EV, Kuligina E, Togo AV, Karpova MB, Ulibina JM, Shutkin VA et al. (2005). Cancer Lett 221 (2):177-183

186. Chiou HL, Wu MF, Chien WP, Cheng YW, Wong RH, Chen CY et al. (2005). Cancer Lett 223 (1):93-101

187. Habalova V, Salagovic J, Kalina I, Stubna J (2005). Neoplasma 52 (5):364-368

188. Osawa Y, Osawa KK, Miyaishi A, Higuchi M, Tsutou A, Matsumura S et al. (2007). Asian Pac J Cancer Prev 8 (1):103-108

189. Cui D, Wang Z, Zhao E, Ma J, Lu W (2011). Lung Cancer 73 (2):153-157

190. Shibuta J, Eto T, Kataoka A, Inoue H, Ueo H, Suzuki T et al. (2001). Am J Gastroenterol 96 (12):3419-3424

191. Jain M, Kumar S, Lal P, Tiwari A, Ghoshal UC, Mittal B (2007). Cancer Invest 25 (5):340-346

192. Akbari MR, Malekzadeh R, Shakeri R, Nasrollahzadeh D, Foumani M, Sun Y et al. (2009). Cancer Res 69 (20):7994-8000

193. Malik MA, Upadhyay R, Modi DR, Zargar SA, Mittal B (2009). Arch Med Res 40 (5):416-423

194. Ladero JM, Agundez JA, Olivera M, Lozano L, Rodriguez-Lescure A, Diaz-Rubio M et al. (2002). Eur J Clin Pharmacol 58 (2):115-118

195. Hong SH, Kim JW, Kim HG, Park IK, Ryoo JW, Lee CH et al. (2006). J Prev Med Public Health 39 (2):135-140

196. Boissy RJ, Watson MA, Umbach DM, Deakin M, Elder J, Strange RC et al. (2000). Int J Cancer 87 (4):507-511

197. Lan Q, Rothman N, Chow WH, Lissowska J, Doll MA, Xiao GH et al. (2003). Cancer Epidemiol Biomarkers Prev 12 (4):384-386

198. Kobayashi M, Otani T, Iwasaki M, Natsukawa S, Shaura K, Koizumi Y et al. (2009). Gastric Cancer 12 (4):198-205

199. Al-Moundhri MS, Al-Kindi M, Al-Nabhani M, Al-Bahrani B, Burney IA, Al-Madhani A et al. (2007). World J Gastroenterol 13 (19):2697-2702

200. Zhang YW, Eom SY, Kim YD, Song YJ, Yun HY, Park JS et al. (2009). Int J Cancer 125 (1):139-145

201. Zhong X, Hui C, Xiao-Ling W, Yan L, Na L (2010). Arch Med Res 41 (4):275-280

202. Frazier ML, O'Donnell FT, Kong S, Gu X, Campos I, Luthra R et al. (2001). Cancer Res 61 (4):1269-1271

203. Lee EJ, Zhao B, Seow-Choen F (1998). Pharmacogenetics 8 (6):513-517

204. Gil JP, Lechner MC (1998). Carcinogenesis 19 (1):37-41

205. Tamer L, Ercan B, Ates NA, Degirmenci U, Unlu A, Ates C et al. (2006). Cell Biochem Funct 24 (2):131-135

206. Kiss I, Nemeth A, Bogner B, Pajkos G, Orsos Z, Sandor J et al. (2004). Anticancer Res 24 (6):3965-3970

207. Katoh T, Boissy R, Nagata N, Kitagawa K, Kuroda Y, Itoh H et al. (2000). Int J Cancer 85 (1):46-49

208. Tiemersma EW, Kampman E, Bueno de Mesquita HB, Bunschoten A, van Schothorst EM, Kok FJ et al. (2002). Cancer Causes Control 13 (4):383-393

209. Butler WJ, Ryan P, Roberts-Thomson IC (2001). J Gastroenterol Hepatol 16 (6):631-635

210. Ye Z, Parry JM (2002). Med Sci Monit 8 (8):CR558-565

211. Yu MW, Pai CI, Yang SY, Hsiao TJ, Chang HC, Lin SM et al. (2000). Gut 47 (5):703-709

212. Gao JP, Huang YD, Lin JA, Zhu QC, Liang JP (2003). Zhonghua Gan Zang Bing Za Zhi 11 (1):20-22

213. Farker K, Schotte U, Scheele J, Hoffmann A (2003). Exp Toxicol Pathol 54 (5-6):387-391

214. Agundez JA, Ladero JM, Olivera M, Lozano L, Fernandez-Arquero M, de laConcha EG et al. (2002). Scand J Gastroenterol 37 (1):99-103

215. Huang YS, Chern HD, Wu JC, Chao Y, Huang YH, Chang FY et al. (2003). Am J Gastroenterol 98 (6):1417-1422
216. Gelatti U, Covolo L, Talamini R, Tagger A, Barbone F, Martelli C et al. (2005). Int J Cancer 115 (2):301-306
217. Imaizumi T, Higaki Y, Hara M, Sakamoto T, Horita M, Mizuta T et al. (2009). Carcinogenesis 30 (10):1729-1734
218. Webster DJ, Flook D, Jenkins J, Hutchings A, Routledge PA (1989). Br J Cancer 60 (2):236-237
219. Ilett KF, Detchon P, Ingram DM, Castleden WM (1990). Cancer Res 50 (20):6649-6651
220. Sardas S, Cok I, Sardas OS, Ilhan O, Karakaya AE (1990). Int J Cancer 46 (6):1138-1139
221. Alberg AJ, Daudt A, Huang HY, Hoffman SC, Comstock GW, Helzlsouer KJ et al. (2004). Cancer Detect Prev 28 (3):187-193
222. Sillanpaa P, Hirvonen A, Kataja V, Eskelinen M, Kosma VM, Uusitupa M et al. (2005). Int J Cancer 114 (4):579-584
223. Ambrosone CB, Kropp S, Yang J, Yao S, Shields PG, Chang-Claude J (2008). Cancer Epidemiol Biomarkers Prev 17 (1):15-26
224. Costa S, Pinto D, Morais A, Vasconcelos A, Oliveira J, Lopes C et al. (2005). Prostate 64 (3):246-252
225. Srivastava DS, Mittal RD (2005). BMC Urol 5:12
226. Hamasaki T, Inatomi H, Katoh T, Aono H, Ikuyama T, Muratani T et al. (2003). Int J Urol 10 (3):167-173
227. Hein DW, Leff MA, Ishibe N, Sinha R, Frazier HA, Doll MA et al. (2002). Environ Mol Mutagen 40 (3):161-167
228. Wadelius M, Autrup JL, Stubbins MJ, Andersson SO, Johansson JE, Wadelius C et al. (1999). Pharmacogenetics 9 (3):333-340
229. Sharma S, Cao X, Wilkens LR, Yamamoto J, Lum-Jones A, Henderson BE et al. (2010). Cancer Epidemiol Biomarkers Prev 19 (7):1866-1870
230. Pasanen M (1999). Adv Drug Deliv Rev 38 (1):81-97
231. Seidegard J, Vorachek WR, Pero RW, Pearson WR (1988). Proc Natl Acad Sci U S A 85 (19):7293-7297

232. Bolt HM, Thier R (2006). Curr Drug Metab 7 (6):613-628
233. Pearson WR, Vorachek WR, Xu SJ, Berger R, Hart I, Vannais D et al. (1993). Am J Hum Genet 53 (1):220-233
234. Zhong S, Wolf CR, Spurr NK (1992). Hum Genet 90 (4):435-439
235. Vos RM, van Welie RT, Peters WH, Evelo CT, Boogaards JJ, Vermeulen NP et al. (1991). Arch Toxicol 65 (2):95-99
236. Wiencke JK, Kelsey KT, Lamela RA, Toscano WA, Jr. (1990). Cancer Res 50 (5):1585-1590
237. Meyer DJ, Coles B, Pemble SE, Gilmore KS, Fraser GM, Ketterer B (1991). Biochem J 274 (Pt 2):409-414
238. Pemble S, Schroeder KR, Spencer SR, Meyer DJ, Hallier E, Bolt HM et al. (1994). Biochem J 300 (Pt 1):271-276
239. Tulsyan S, Chaturvedi P, Agarwal G, Lal P, Agrawal S, Mittal RD et al. (2013). Mol Diagn Ther
240. Kiran B, Karkucak M, Ozan H, Yakut T, Ozerkan K, Sag S et al. (2010). J Gynecol Oncol 21 (3):169-173
241. Kwon DD, Lee JW, Han DY, Seo IY, Park SC, Jeong HJ et al. (2011). Korean J Urol 52 (4):247-252
242. Sreeja L, Syamala V, Hariharan S, Syamala VS, Raveendran PB, Sivanandan CD et al. (2008). J Exp Ther Oncol 7 (1):73-85
243. Hohaus S, Mansueto G, Massini G, D'Alo F, Giachelia M, Martini M et al. (2007). Leuk Lymphoma 48 (3):564-569
244. Mossallam GI, Abdel Hamid TM, Samra MA (2006). J Egypt Natl Canc Inst 18 (3):264-273
245. Scott RB, Matin S, Hamilton SC (1990). J Lab Clin Med 116 (5):674-680
246. Coles BF, Kadlubar FF (2003). Biofactors 17 (1-4):115-130
247. Bai YL, Zhou B, Jing XY, Zhang B, Huo XQ, Ma C et al. (2012). Asian Pac J Cancer Prev 13 (10):5019-5022
248. Matakova T, Sivonova M, Halasova E, Mistuna D, Dzian A, Berzinec P et al. (2009). Eur J Med Res 14 Suppl 4:275-279
249. Dobrossy L (2002). Lancet Oncol 3 (6):374-381

250. Hernandez BY, Green MD, Cassel KD, Pobutsky AM, Vu V, Wilkens LR (2010). Hawaii medical journal 69 (9):223-224

251. Calderon-Garciduenas AL, Rivera-Prieto RA, Ortiz-Lopez R, Rivas F, Barrera-Saldana HA, Penaloza-Espinosa RI et al. (2008). Am J Hum Biol 20 (2):191-193

252. Lazcano-Ponce EC, Miquel JF, Munoz N, Herrero R, Ferrecio C, Wistuba, II et al. (2001). CA Cancer J Clin 51 (6):349-364

253. Weiss SE, Tartter PI, Ahmed S, Brower ST, Brusco C, Bossolt K et al. (1995). Cancer 76 (2):268-274

254. Millikan R, DeVoto E, Newman B, Savitz D (1995). Breast Cancer Res Treat 35 (1):79-89

255. Ellard GA (1976). Clin Pharmacol Ther 19 (5 Pt 2):610-625

256. Hildebrand M, Seifert W (1989). Eur J Clin Pharmacol 37 (5):525-526